4th Grade Science Volume 4

© 2013 Todd Deluca
OnBoard Academics, Inc
Newburyport, MA 01950

800-596-3175
www.onboardacademics.com

Table of Contents

The Skin

What percent of your body weight is made up of skin?

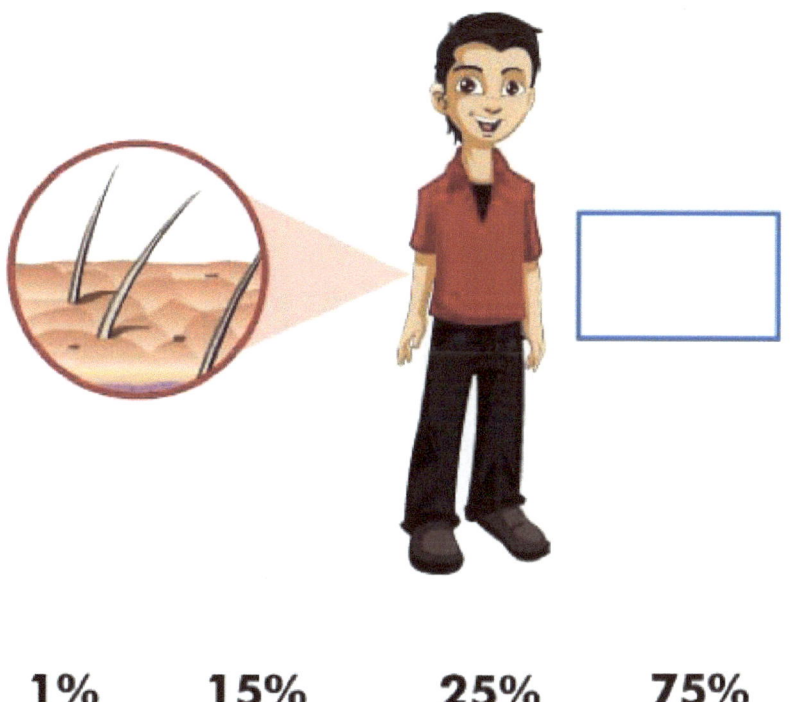

1%   **15%**   **25%**   **75%**

Your skin is the largest organ in your body and accounts for 15% of your body weight. This means that an adult who weighs 160 pounds has 24 pounds of skin!

The Three Layers of Skin

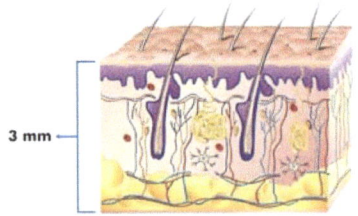

Your skin is about 3mm thick and is made up of three different layers.

The outer layer, the layer that you can see, is called the epidermis and is made up of layers of cells. You have more layers where your skin is the thickest like the palms of your hands and the soles of your feet. You are continuously loosing the outer layer as dead skin falls off your body. Every minute you lose about 30,000 dead skin cells. The good news is that new cells from the bottom of the epidermis push up to replace these dead skin cells.

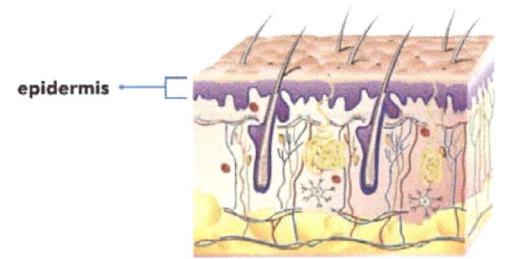

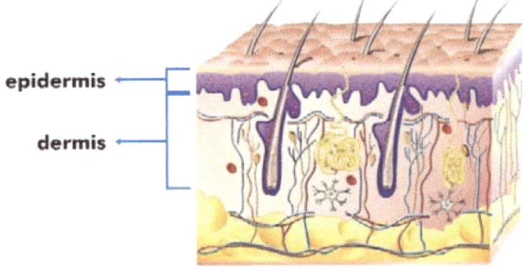

The next layer called the dermis is where a lot of the action within your skin takes place. There dermis contains many blood vessels, nerves cells, and oil and sweat glands. The roots of your hair are also found in the dermis.

Below the dermis is a layer of fatty tissue called the subcutaneous layer. This layer helps to keep your body warm and acts as a cushion to help protect you from injury.

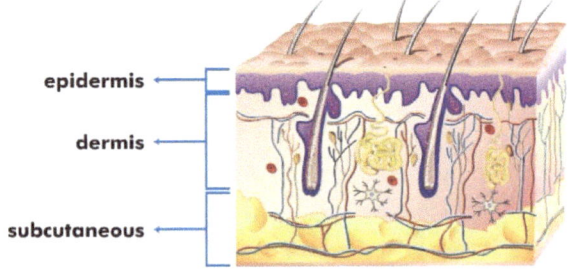

In which layer of the skin are these found?

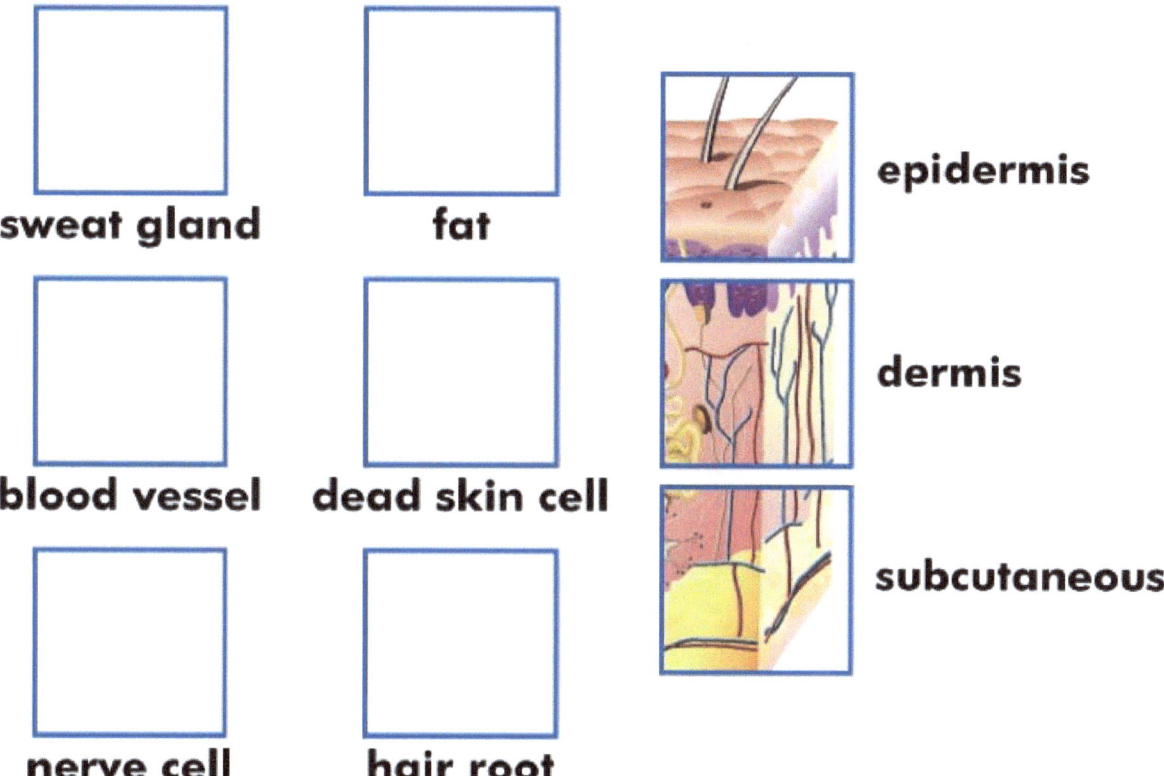

sweat gland

fat

epidermis

blood vessel

dead skin cell

dermis

nerve cell

hair root

subcutaneous

Your skin in like a suit of armor.

Your skin acts like a suit of armor for your body.

It keeps dirt, bacteria and other harmful organisms from getting in and water and food from seeping out.

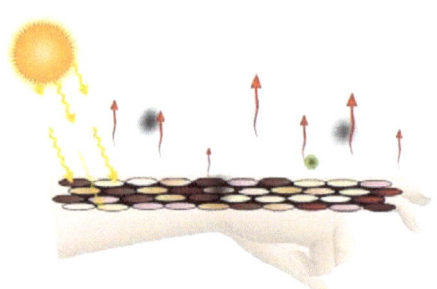

It also contains an important pigment called melanin that keeps out harmful radiation from the sun. If you didn't have any melanin you'd shrivel up.

> **Your skin acts like a suit of armor for your body. It keeps harmful substances from getting in, and it stops fluids from leaking out. It also protects you from the Sun's harmful rays.**

People have different colors of skin because of the levels of melanin. Draw a circle around the arm that most matches your color of skin.

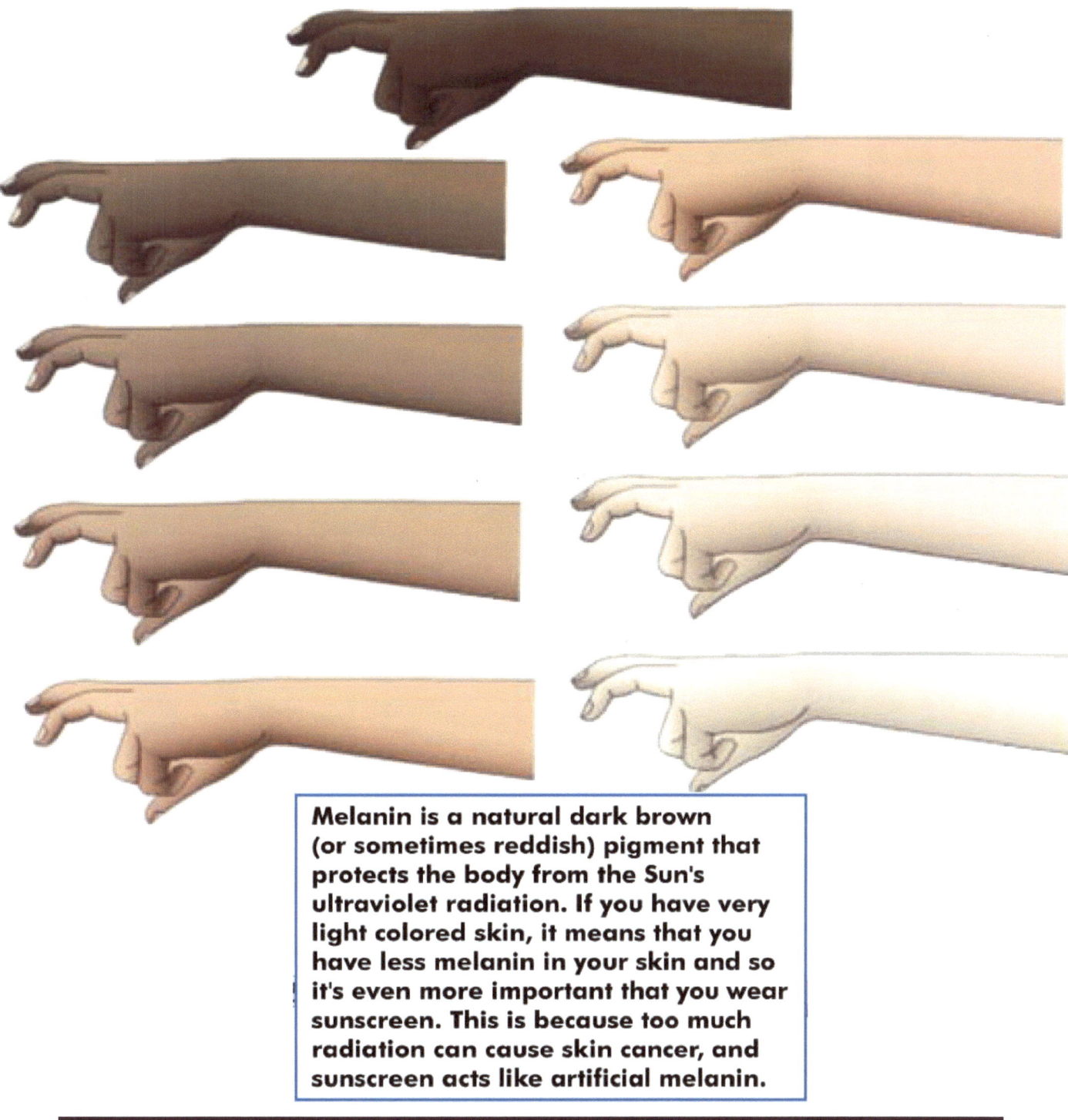

Melanin is a natural dark brown (or sometimes reddish) pigment that protects the body from the Sun's ultraviolet radiation. If you have very light colored skin, it means that you have less melanin in your skin and so it's even more important that you wear sunscreen. This is because too much radiation can cause skin cancer, and sunscreen acts like artificial melanin.

Your skin is a sensing organ and communicates with your brain.

If your hand were to touch these items what would it communicate to your brain?

Write your answers below each item, if you have colors you can make a colored dot to correspond with the message.

 PRESSURE

 HEAT

 COLD

 PAIN

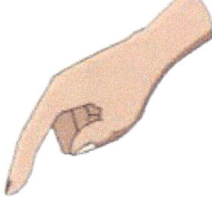

   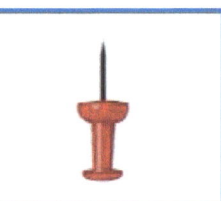

Your skin acts as a thermostat for your body.

Match the temperature to the skin condition.

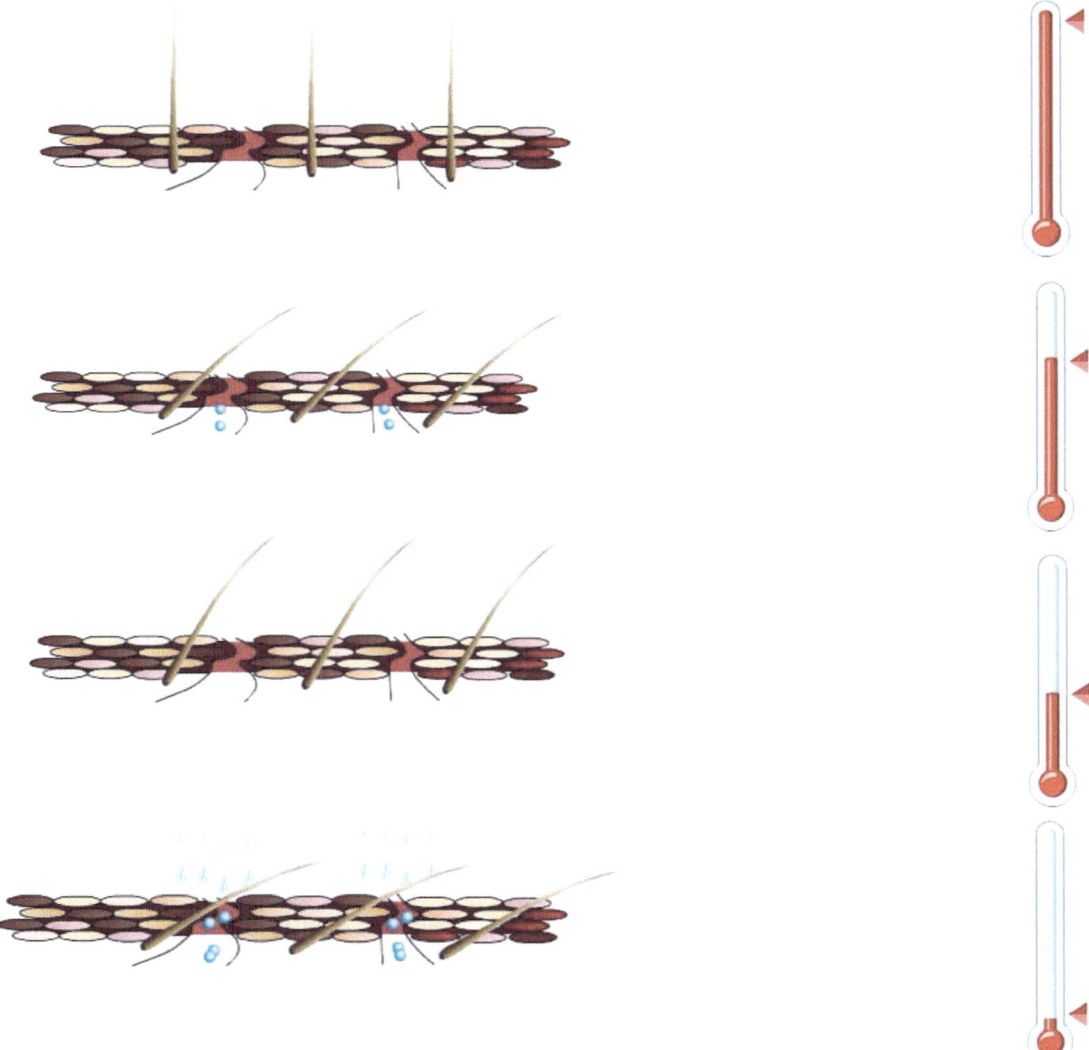

Your skin acts as a thermostat for your body.
When it's hot, you sweat. Sweating cools
your body because when water evaporates
it uses up heat from the surface of the skin.
When your body is cold, the hair on your skin
stands up to trap warm air around your body.

Skin Facts; True or false?

Sweating helps to remove heat from your skin and helps to keep you cool. _____

The epidermis is the outer layer of the skin. _____

Melanin is a type of nerve cell. _____

You lose about 30 skin cells every minute. _____

Your skin has several different types of nerve cells. _____

The more melanin you have, the darker your skin. _____

The Skin Quiz

1. The thickness of our skin is _____.
    a. 3 mm
    b. 3 cm
    c. 30 mm
    d. 3 m

2. The outermost layer of the skin is called the

    _____.

3. Sweat glands are present in the epidermis. True or false?

4. The skin has pigment called _____ which protects us from the harmful radiation of the Sun.
    a. melamine
    b. melanin
    c. rhodopsin
    d. iodopsin

5. The skin acts as a thermostat for our body. True or false?

The Brain

The Brain

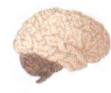

The human brain is the body's control center. The brain is a hugely complex organ that enables us to think, remember, dream, move, feel, see hear, taste, and smell.

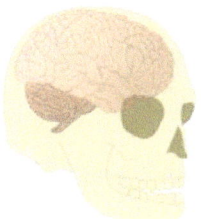

Located inside the skull for protection, the brain communicates with all parts of the body via the spinal chord a length of nerve tissue that extends through the backbone and branches out to all other parts of the body. The brain receives signals called nerve impulses from different parts of the body through the spinal chord and sends out its instructions via the same route.

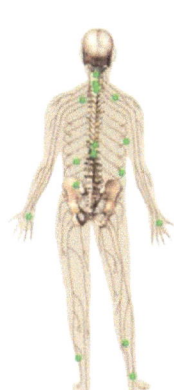

Some of these instructions are voluntary such as the instruction to raise our arm. Other instructions are involuntary such as breathing, our heart beat or digestion.

Pinkish gray in appearance and wrinkled looking due to its many folds, the brain has three main parts; the cerebrum, the cerebellum and the brain stem. Label the three parts of the brain on the illustration below. Read the rest of the passage for help

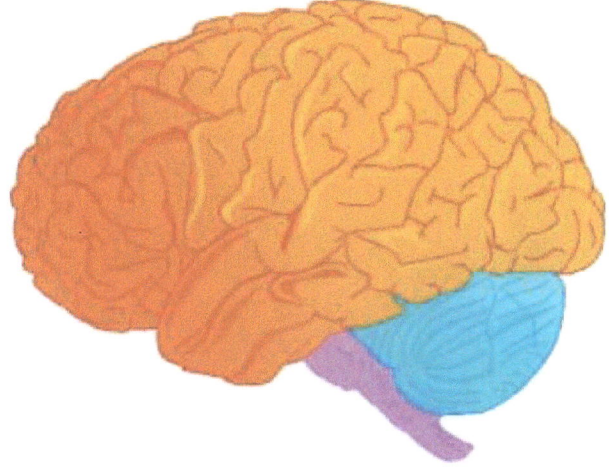

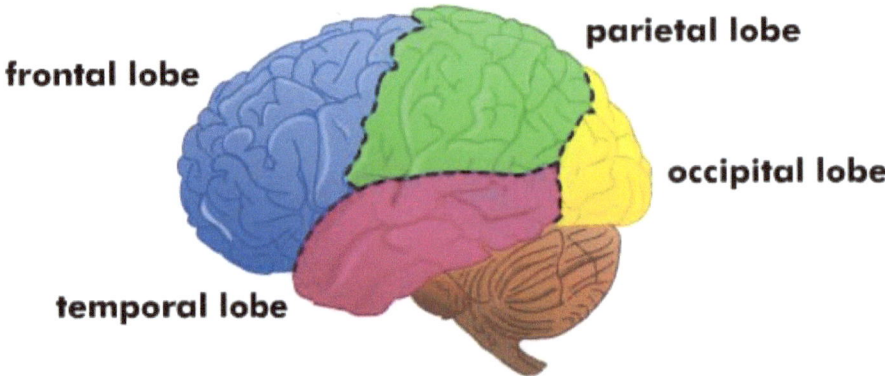

The cerebrum is the largest part of the brain and has many functions such as thoughts, language and comprehension. The cerebrum is divided into four sections called lobes and each lobe has a specific function.

The frontal lobe is responsible for planning, problem solving, and speech. The temporal love deals with hearing and memory.

The parietal lobe helps us to interpret touch, pressure, temperature and pain.

The occipital lobe deals with vision

At the back of the brain under the cerebrum is the cerebellum. The cerebellum coordinates body movement, posture and balance.

The brain stem connects the brain with the spinal chord and is located in front of the cerebellum.

The lower part of the brain stem is called the medulla and is responsible for controlling involuntary processes in your body such as breathing.

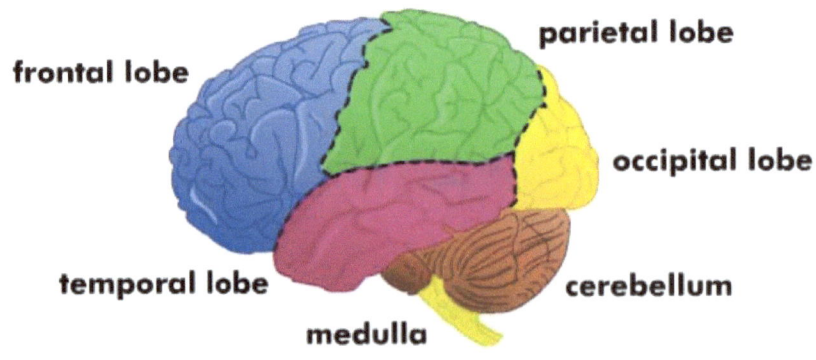

Did you label the brain's three main section like this?

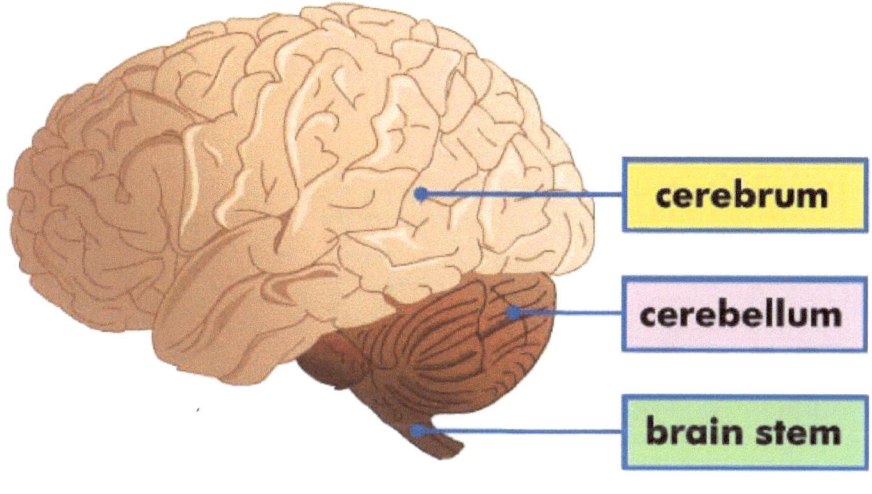

Now Label the lobes of the brain.

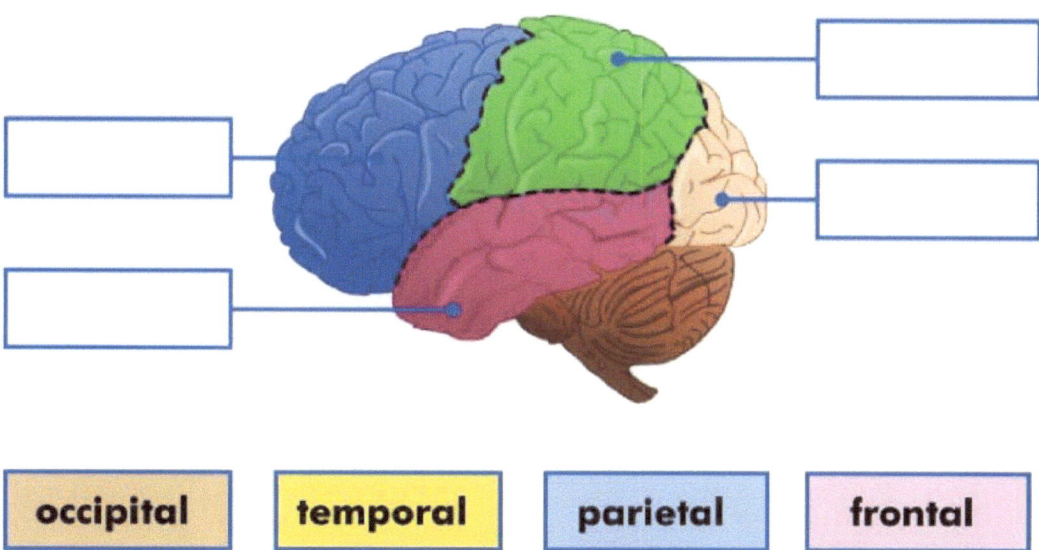

| occipital | temporal | parietal | frontal |

Which lobe am I using?

I just tripped on a rock and stubbed my toe!	
It's very painful.	
It reminds me of when I got hurt playing soccer last year.	
I can see blood coming out of my sock.	
I have a scarf that I can use as a bandage until help arrives	
Oh good, I can hear someone coming.	

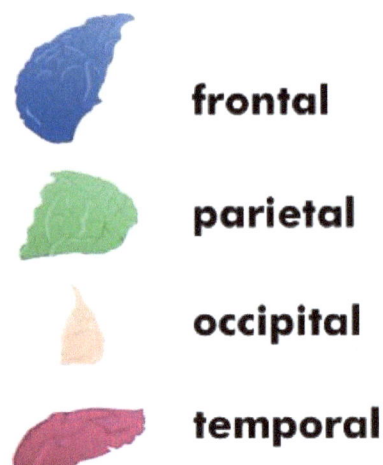

frontal

parietal

occipital

temporal

Are you left brained or right brained?

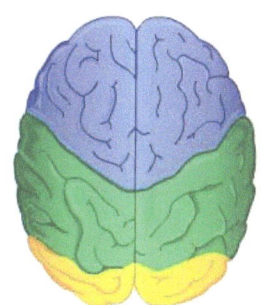

The brain is actually divided into two halves know as hemispheres.

Although its a bit confusing, the right side of the brain controls the left side of the body and the left side of the brain controls the right side of the body.

Some scientists believe that people have an orientation toward one side of their brain meaning one side of the brain is more influential than the other side.

With this theory they believe that left brained people are better at math and science while right brained people are better at art and music.

If that's true, do you thin your are left or right brained? _____

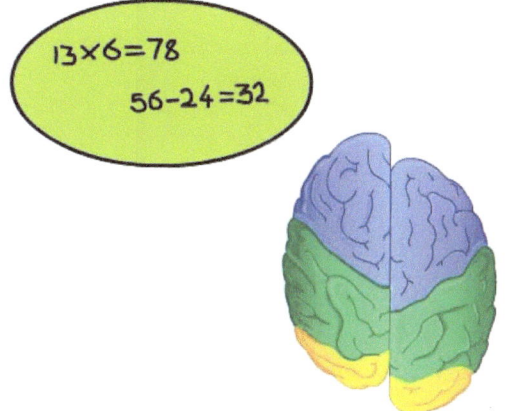

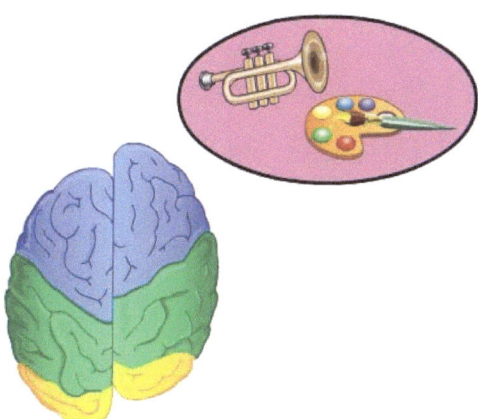

Name: _____

The Brain Quiz

1. The brain communicates with all parts of the body via the _____.
    a. skull
    b. blood vessels
    c. spinal cord

2. Digestion is a voluntary instruction given by the brain. True or false?

3. The color of the brain is _____.
    a. pinkish gray
    b. bluish gray
    c. reddish gray

4. The _____ of the brain is responsible for thought, language, and comprehension.
    a. cerebellum
    b. medulla
    c. cerebrum

5. The frontal lobe in the cerebrum is responsible for temperature and pain. True or false?

The Eye

Which frog has the bigger mouth.

Answer by observing the two frogs and then measure each frogs mouth and answer.

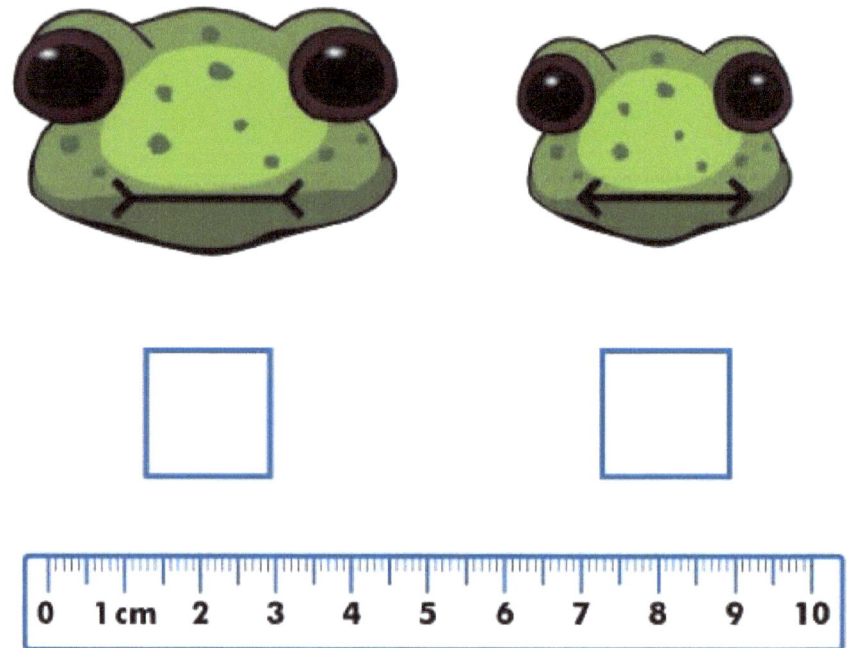

When we see something that isn't really there, we call this an optical illusion. However, it's our brains rather than our eyes that have been fooled.

www.onboardacademics.com

How do we see?

When we look at an object, light bounces off of the object and into our eyes. If it is completely dark, no reflective light enters our eyes and that is why we don't see anything.

When light reaches us it pass through a clear outer layer of the eye called the cornea. It then travels through the dark hole in the middle of the eye called the pupil. The colored part of the eye that surrounds the pupil is called the iris. Muscles of the iris control the amount of light that enters the eye by changing the size of the pupil. Light enters the pupil and reaches the lens. The job of the lens is similar to a lens in a camera, it helps to focus the image and to make it sharp and clear. The focused image then appears on a sensitive lining on the back of our eye called the retina. The retina contains special cells called cones and rods. Cones help us to see in bright light and to distinguish color while rods help us to see in dim light.

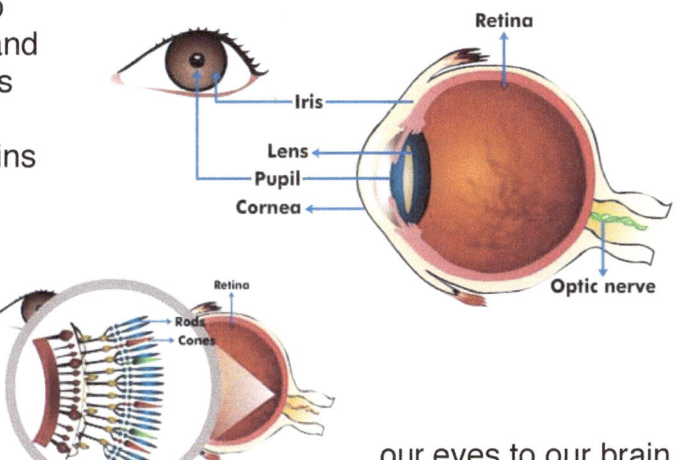

The retina also has many nerve endings that send the information about the light that has entered This information travels through

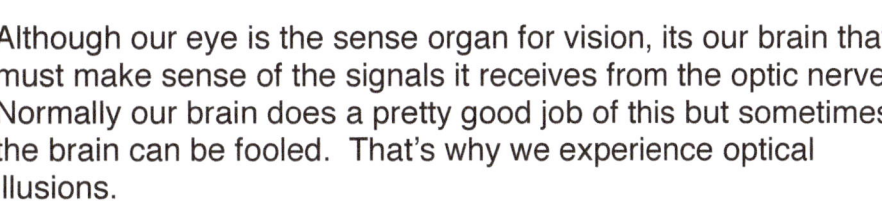

our eyes to our brain. the optic nerve.

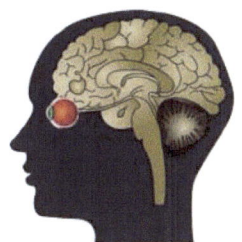

Although our eye is the sense organ for vision, its our brain that must make sense of the signals it receives from the optic nerve. Normally our brain does a pretty good job of this but sometimes the brain can be fooled. That's why we experience optical illusions.

Label the parts of the eye.

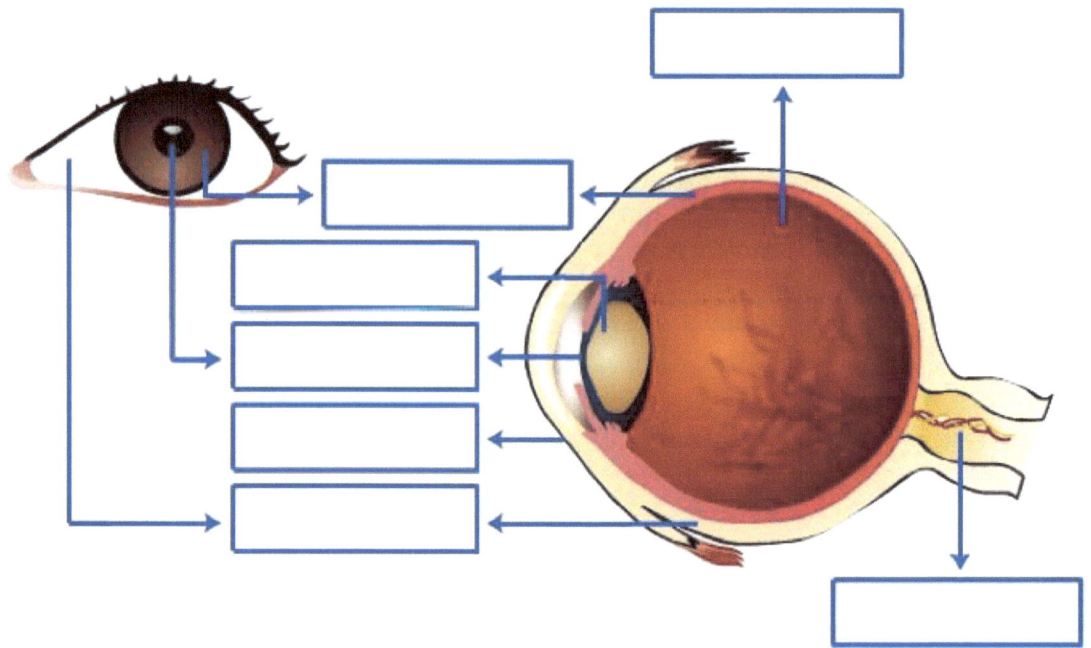

Retina   Iris   Lens   Pupil   Cornea   Sclera   Optic nerve

What part of the eye am I?

Match the eye part to the correct description. You may have to use the process of elimination to learn a new term.

I carry information from the retina to the brain.	**pupil**
I'm the colored part of the eye that controls the amount of light that enters the pupil.	**optic nerve**
I'm a clear disk that helps to focus an image.	**lens**
I am the clear outer covering of the eye.	**retina**
I'm the sensitive lining at the back of the eye.	**cornea**
I'm the hole at the center of the iris through which light passes.	**iris**
I'm the tough white outer layer of the eye.	**sclera**

Find your blind spot.

Each of your eyes has a blind spot. This is the point at which your optic nerve connects to your retina. It is called the blind spot because when an image lands o that part of our retina, you can't see it. This is because there aren't any rods or cones at this point.

To test your blind spot, place your book or computer upright and stand about three paces away. Close your right eye and with your left eye look at the +. Walk towards the image slowly. At a certain distance the dot on the left side will disappear. This is the point at which the image of the dot falls on the blind spot of the retina of your left eye. To test the blind spot on your right eye, close your left eye and look at the dot with your right eye. As you approach your booklet, the + will disappear.

Create an optical illusion.

Put your index fingers to-gether like this about 30 cm in front of you at eye level. Then look at the wall behind your fingers. What do you see?

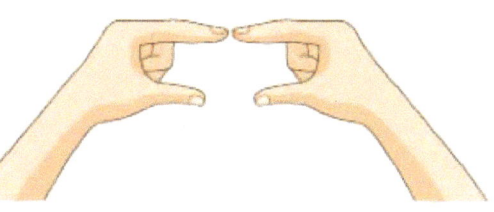

You should see an extra finger. Why do you think this happens?

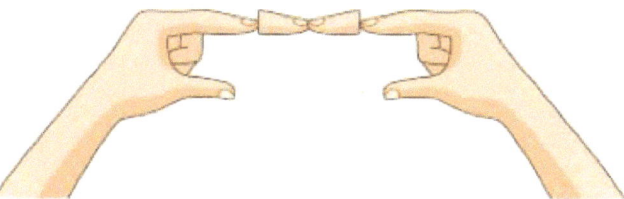

Name: _____

The Eye Quiz

1. An _____ is seeing something which is not really present.
    - a. optical contraction
    - b. optical transformation
    - c. optical illusion
    - d. optical theory

2. The clear outer covering of the eye is called the _____.
    - a. iris
    - b. pupil
    - c. retina
    - d. cornea

3. Our eyes have a lens which helps us focus images. True or false?

4. The tough white outer layer of the eye is called the _____.
    - a. pupil
    - b. sclera
    - c. retina
    - d. iris

5. Our pupils dilate when there is less light. True or false?

The Nervous System

The Job of the Nervous System

allow us to sense

Signals from sensory organs such as our skin and our ears are transported by our nervous system to our brain and and interpreted as a sense such as sound, a texture or pressure.

voluntary activity

Our nervous system carries commands from our brain that result in voluntary movement such as movement. For example if our nervous system transports an inbound signal to our brain that a nasty dog is coming our way, our brain sends an outbound signal to get going.

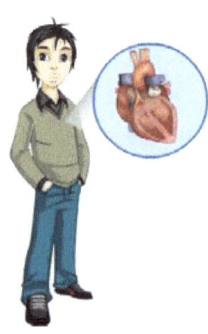

involuntary activity

Our nervous system is also responsible for involuntary activity in the body such as the beating of our heart and secretions from glands. We call this the automatic nervous system because we don't even think about these activities.

The nervous system is the body's command and control system. It allows us to sense, controls voluntary movement (such as talking or walking), and also involuntary movement such as our heart beat. The brain is the command center of the nervous system.

The Division of the Nervous System

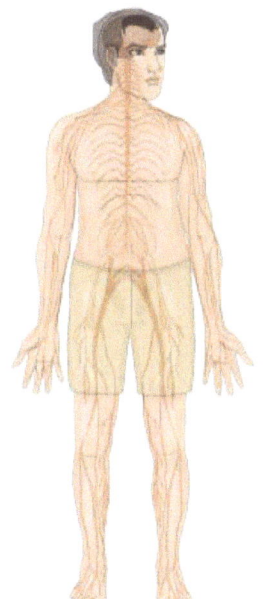

There are two divisions or main parts of the nervous system, the central nervous system which consists of the brain and spinal cord, and the peripheral nervous system which is a network of nerves that connects the central nervous system to all parts of the body.

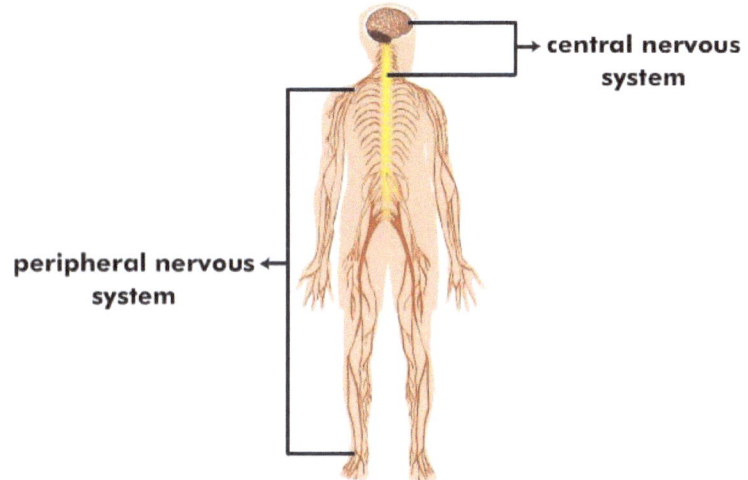

central nervous system

peripheral nervous system

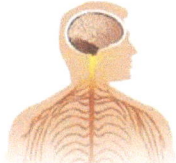

Specifically this network of nerves in the peripheral system is made up of 12 pairs of cranial nerves from your brain and 31 pairs of spinal nerves from your spinal cord.

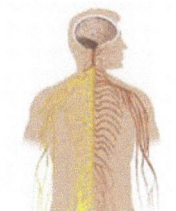

For protection the brain is located in the skull and the spinal cord inside a tunnel in the back bone.

The spinal cord is about 45 cm long in adult males and 43 cm long in adult females. It extends from the base of your brain to the area that corresponds with the small of your back.

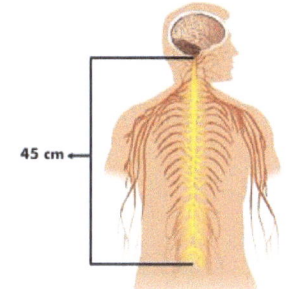

45 cm

There are two divisions of the nervous system: the central nervous system, which consists of the brain and the spinal chord, and the peripheral nervous system which consists of a network of nerves that connect the central nervous system to all parts of the body.

How does the nervous system work?

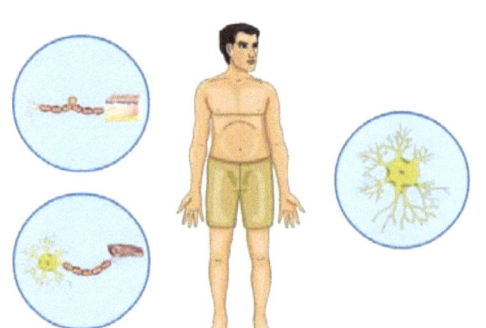

To understand how the nervous system works we need to understand the function of three different types of cells within the nervous system; sensory neurons, motor neurons and interneurons.

The role of sensory neurons is to collect and pass on information about the outside world that we get from our sensory organs. For example, if you touch something with your hand, receptors in your

skin that are attached to sensory neurons will capture and pass on that touch information in the form of an electro chemical signal that we call a nerve impulse. The nerve impulse travels from your hand up your arm to the spinal cord and into your brain.

When the impulse enters your brain and different neurons called interneurons will try to make sense of the information and then send back an outgoing message. For example your brain may send out instructions to pick up the object for closer examination.

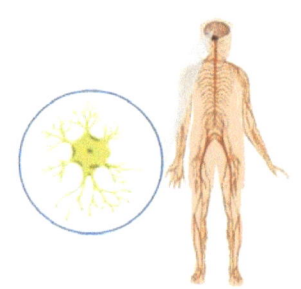

These instructions will travel back on the same path in reverse direction and will reach the skeletal muscles in your hand by way of the third type of nerve cell called motor neurons. Motor neurons control the movement and activity of muscles and glands.

When there isn't time for the brain to be involved in the decision making process; for example if you touch an object that is red hot, the information travels directly from the sensory neurons to interneurons in the spinal cord and

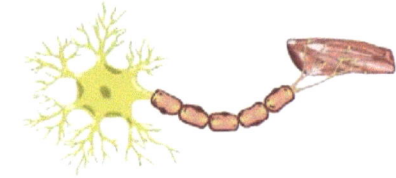

then directly back to the motor neurons in our hand. The result is that you move your hand away immediately.

A moment later the information from the pain receptor in your had will reach your brain and you will experience the sense of pain. Ouch!

Review: Label the parts of the brain.

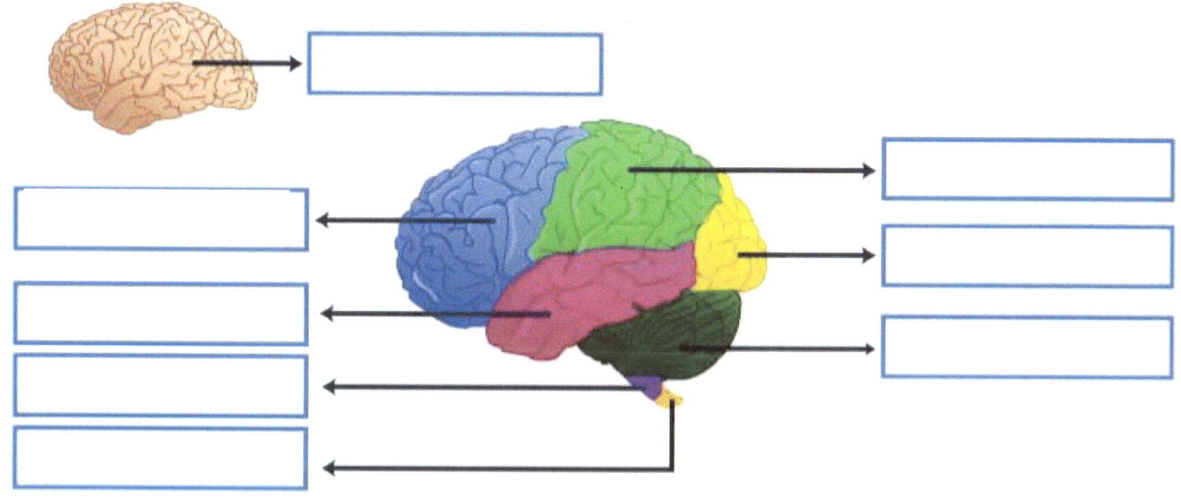

frontal lobe   parietal lobe   spinal cord   cerebellum

temporal lobe   occipital lobe   brain stem   cerebrum

The Nervous System Quiz

1. The _____ is/are the command center of the nervous system.
    a. heart
    b. brain
    c. kidney
    d. nerves

2. The nervous system controls both voluntary and involuntary movements. True or false?

3. The _____ consists of a network of nerves that connect he central nervous system to all parts of the body.
    a. central nervous system
    b. peripheral nervous system

4. _____ neurons are responsible for collecting information from the environment outside the body.
    a. inter
    b. sensory
    c. connect
    d. motor

5. The _____ is the largest part of the brain.

Newburyport, MA 01950

1-800-596-3175

OnBoard Academics employs teachers to make lessons for teachers! We create and publish a wide range of aligned lessons in math, science and ELA for use on most EdTech devices including whiteboard, tablets, computers and pdfs for printing.

All of our lessons are aligned to the common core, the Next Generation Science Standards and all state standards.

If you like our products please visit our website for information on individual lessons, teachers licenses, building licenses, district licenses and subscriptions.

Thank you for using OnBoard Academic products.